©2019 Creation Science Evangelism
This book may be copied whole, not in part, for free distribution only.
Published 2019
ISBN978-1-7335128-5-5
Printed in the United States of America
Artist: Sonja Taljaard

Creation Science Evangelism, Inc.
is a part of Dr. Dino Creation Ministry
488 Pearl Lane
Repton, AL 36475
1.855.BIG.DINO (244.3466)
www.drdino.com
drdino@drdino.com

Introduction

Grandpa Hovind, that's me, has five grandchildren!

The world needs the hearts of the fathers turned to their children *(Malachi 4:6 "And he shall turn the heart of the fathers to the children, and the heart of the children to their fathers, lest I come and smite the earth with a curse." Luke 1:17 " . . . and he shall go before him in the spirit and power of Elias, to turn the hearts of the fathers to the children, and the disobedient to the wisdom of the just; to make ready a people prepared for the Lord.")* Character building stories are a great way to do that. Nearly every culture on earth passes on its collective wisdom, morals, and knowledge to the next generation via stories, parables, and fables.

Come along with me and my grandchildren as we explore God's amazing creation. We will learn valuable life lessons, studying various animals, plants, rocks, and the earth's geography.

It is my prayer that these stories will draw you, your children, and grandchildren closer to the great God of the universe.

Kent Hovind

Ants

Dear Grandkids,

I love you s-o-o-o-o much! I want you all to grow up, love Jesus, and help others find Him. I also want you to love the beautiful world God made.

King David, the one who killed Goliath when he was a boy, had a son named Solomon. God made Solomon the smartest man who ever lived. Solomon said we should study the ants. Let's do it. Let's study the ants! I want you to find some ants and watch them for a few minutes. All the worker ants are girls. How did Solomon know to say "consider *her* ways and be wise?" When you watch ants you can learn lots of really good lessons.

Lesson #1 Ants are always busy working while they can. They know winter is coming when it will be too cold to work, so they work hard to gather food and save it up. We need to do the same thing. Learn to work hard and save up for when we can't work.

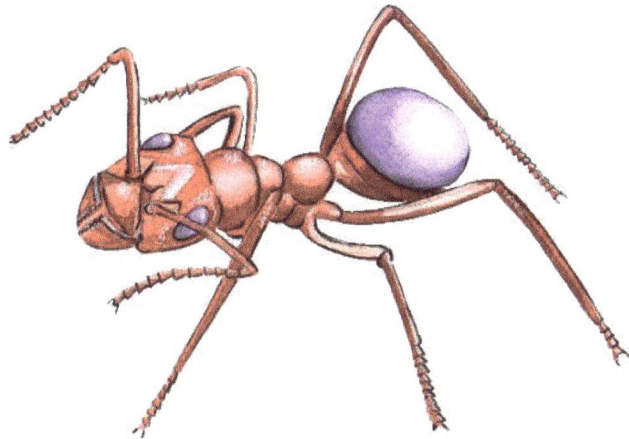

Lesson #2 Throw a few crumbs down for the ants and watch. *They never argue about who gets the biggest piece.* That's a good lesson to learn to please Jesus!

Lesson #3 Nobody tells the ants to get to work. When they see something that needs to be done, they just do it. Do you pick up your toys or put dishes in the sink or help Mom and Dad *without being told?*

Lesson #4 When I watched the ants, I noticed that they all help take care of the babies. When it rains and water goes into their nest, or if someone steps on their ant hill, they all hurry to carry the baby ants to safety. When there is danger, be sure to watch out for the littlest ones first.

Lesson #5 If anyone disturbs their nest, they all come out to defend it. If anyone tries to hurt one of your family members, you should stand up for your family. This means your Christian family also. That's why ants bite you if you get too close. They think you are going to hurt them.

What other lessons can we learn from the ants? They don't get mad when other ants bump them or even if another ant steps on them! See what else you can learn from the ants!

I love you! Grandpa

Armadillo

When I lived in Longview, TX, we had a house in the woods that I built. It had a creek. There was an armadillo that lived in the woods behind our house. Every night he would walk around our yard digging in the ground looking for worms, bugs, roots, and fruits.

God is s-o-o-o-o smart! He made the armadillo with real good hearing and smell, but he can't see very well since he sleeps all day and is up all night. When an enemy tries to hurt him, he curls up in a ball and has armor plates to protect himself.

Armadillos can swallow a lot of air so they float really well and can swim across rivers and lakes. They are shy and run away from people if they can. They can have anywhere from four to twelve babies at the same time! The babies will either be all girls or all boys. How would you like to have eleven sisters exactly like you?!

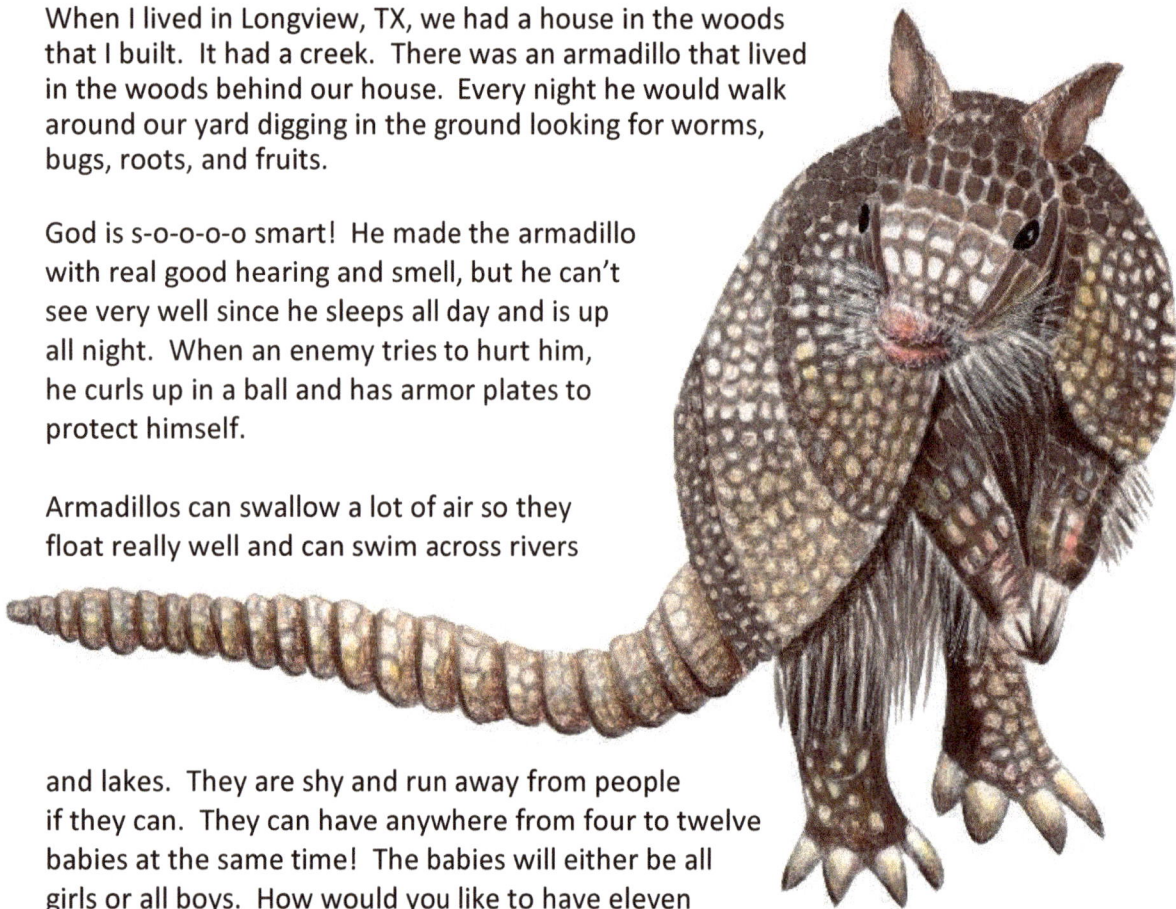

Some kinds of armadillos are only six inches long when they grow up and some kinds are five feet long! The one that lived in our yard was about two feet plus a tail that was 1 & 2/3 feet long. We threw food out at night so he would come eat it.

Armadillos have strong claws and can dig a hole in the dirt real fast! They dig holes under trees for their house.

We will go see one someday. Always love Jesus and study His amazing world!

I love you,
Grandpa

Bats

Dear Grandkids,

I pray for you a lot. I want you to be hard workers and love Jesus.

Last night as I came out of church, one of the men said, "Hey, there's a bat on the table!" Two bats were in the air fighting over a bug and the big one knocked the little one down to the table. He was curled up in a ball about as big as a golf ball. He only weighed as much as a letter.

God is s-o-o-o-o smart! The bat was called "the little brown bat." There are about 1,100 different kinds of bats found all over the world. They look like a mouse with wings. The biggest bat in the world has wings that reach five feet!

The one I saw last night only had a ten-inch wingspan. I slid a piece of paper under him to pick him up and carry him to a tree in a dark corner so he could rest up from being hurt. You must be careful to never pick up a bat with your hands. If you do touch a bat, wear very thick gloves. Bats have razor-sharp teeth like needles! They don't know you are trying to help them, so they will bite you to protect themselves.

Lots of bats have a bad disease called rabies. If a bat has rabies and bites you, he will give you the rabies disease. That would make you very, very, very sick. Some people die from rabies.

Bats live about ten years. They love to live in caves or inside holes in your roof or in trees. They can catch bugs right out of the air! The little brown bat eats one-half of his body weight in insects every night! That's like Matthew eating 60 hamburgers in one day!! Bats sleep all day and fly around hunting bugs all night. Sometimes you can see them fly around right after it gets dark. They can see, but at night they make a squeak sound to find the bugs. Their big ears hear very well. It's called echo location.

In the winter, they hibernate. That means they sleep for weeks at a time. The bats sleep upside down, hanging from the roofs of caves. After bats have lived in a cave for hundreds of years, the bat do-do, called guano, gets really thick on the floor. Sometimes over one hundred feet thick! Farmers love to get guano to put on their fields for fertilizer to make the crops grow faster and bigger.

Bats are really fun to study. God really made an amazing animal when He made the bat. Always love God and tell others about Him!

Bird Feathers

God thinks of everything! He made birds with three kinds of feathers. They have a real fluffy feather when they are babies.

These feathers won't let the birds fly, but they do keep it really warm and provide protection from bumps.

As they grow bigger they get contour feathers. They help the bird's body be smooth so they can fly faster and look good. The shaft is in the middle with the same amount of feathers on each side.

The other kind of feather is a flight feather. They are bigger, stronger, and have the shaft off to one side.

All feathers are made out of a very special material that is really strong, really light, and really pretty. Your hair and fingernails are made out of the same stuff.

God made this amazing world in only 6 days. I can't wait to see the house He has made for us in Heaven! He's been making it for a long time. Be sure you have Jesus in your heart so you can be with Jesus in Heaven forever.

Bluebirds

Today I saw a beautiful bluebird! Not a blue jay. That's a bigger bird with blue feathers. Bluebirds have light blue feathers on top and light orange feathers on their chest and belly. He was so pretty!

They love to build their nests in holes in trees or even in mailboxes!

Bird feathers are amazing! God is s-o-o-o smart! Birds can point their tail and wing toward the ground when it rains and the water runs off like a rain coat. They don't even get wet inside.

Some of their feathers are really soft, called down, just to keep them warm. Others are used for flying and are very strong.

The parts of the feather "zip" together like a zipper or Velcro. If you pull gently, you can see the hooks. Birds always pull their feathers through their mouth to zip them back together (like the baggies that have zip lock tops or a zipper on your jacket).

It's a good thing God gave the bluebirds a very flexible neck! Can you lick your own back? Birds can!

I hope you always love Jesus and love studying His amazing creation. I can't wait to go to heaven and see the huge house He is building for us!

I love you!

Bugs

Today a tiny, tiny bug landed on my ear. That little bug had eyes, a mouth, six legs, wings, a shell, two antennae, a stomach and six feet. Imagine how smart God is to make such tiny things that work perfectly! Imagine how small his eyelashes are!

Last week I caught a bug only as big as a dot, but he got away. I love studying what God has made. I can't wait to get to Heaven and see all the stuff He has been preparing in our new house. That is going to be awesome!

God must like bugs because He made thousands and thousands of different kinds. Some kinds of bugs hurt people but most do not. Some people don't want to believe God made everything. They think bugs just slowly evolved out of dirt! Those people need someone to tell them about Jesus. That's what we do at Dinosaur Adventure Land.

I hope you kids all grow up to help us work for Jesus at DAL!

Butterflies and Moths

God's world is so fun to study. I love it. God made so many kinds of bugs that you could study them all your life and never be able to study all of them; there are way too many.

I really like studying butterflies. Most of them are beautiful. Butterflies and moths look alike, but they are different in three ways. Most moths fly at night while butterflies fly in the daytime.

Moths rest with their wings flat while butterflies fold their wings up to rest.

Most moths have antennae that look like little feathers while butterflies have antennae that look like a slender hair with a knob on the end.

Did you know that butterflies and moths hear and smell with their antennae! They taste plants with their feet!

When they hatch out of their eggs, they look like a hairy worm. This hairy worm is called a caterpillar. They eat a lot of leaves and grow bigger. When they get big enough, they spin a cover that looks like a sleeping bag. It is called a cocoon.

They stay in their cocoon for a few weeks where a miracle happens. They turn from a worm into a butterfly! Where did the wings come from? Only God can do that! I have a clock somewhere made out of beautiful butterfly wings from Brazil.

Butterflies and moths cannot bite you, but some of them have real tiny claws on their feet that might pinch you just a little bit if they land on your skin.

Maybe your dad could help you catch some pretty butterflies to look at. Don't touch their wings, or they won't be able to fly anymore.

I love you!

Always love Jesus and study the amazing world He made.

Buzzards – Turkey Vultures

God is s-o-o-o-o-o smart! He made amazing animals for us to study. Today I saw six huge birds flying in a really slow circle. The real name of these birds is Turkey Vultures, but everyone calls them Buzzards.

They are huge! They can reach their wings out six feet! If one was standing in your kitchen, he could put his head on the kitchen table! But you would not want him in your kitchen! The only thing he eats is dead, rotten animals. They really stink badly!

It's a good thing God made buzzards to eat animals that die in the woods or get run over on the road. If dead animals were not eaten by ants or buzzards, the dead animals would stink up the whole neighborhood.

Buzzards have mostly black feathers but they don't have any feathers on their head. If they did, the blood from the animals they eat would get stuck in their feathers. God thinks of everything! They look pretty ugly but they won't hurt you.

God gave them really good eyes. They can fly way up in the sky and see a small dead animal.

The buzzards know how to find places where the hot air is blowing up. When they have their wings spread out, the blowing hot air lifts them up. They can soar without flapping their wings for a long time that way.

Sometimes they fly so high you can barely see them, but they can see you! God is like that with us. Even though we can't see Him, He sees us!

Keep watching the sky to see who can spot a buzzard first.

Goslings

Dear Grandkids,

We have four huge Canadian geese that live in a small pond here. Today I saw two of them with five babies walking around the grass!

The babies are called goslings. They have really fluffy brown feathers to keep them warm. They don't need stiff flight feathers since they can't fly. God always gives us what we need *when* we need it. He is s-o-o-o-o smart!

The goslings were following their mom and dad really close to stay safe and to learn how to be a goose. They were learning what to eat and how to watch out for enemies that might hurt them.

Their nest is lined with very soft feathers. The babies get under mom and dad's wings to sleep. That keeps them safe!

God says He wants us to stay under His wings to be safe. All He wants for us is what is best.

Always love Him and obey Him even if you don't understand.

Hawks

Today I was out walking, and a hawk flew down and landed on the wall.

There are lots of kinds of hawks. This one was about the size of a crow.

A hawk's feathers are light brown on their belly, and dark brown or spotted on their back and wings. The one I saw was a red-shouldered hawk. He has this name because of the patch of light brownish-red feathers on his shoulders.

Hawks eat mice, snakes, fish, squirrels, lizards, and insects. They have very sharp claws and beaks. Hawks are good to have around farms because they eat the rats and mice that try to eat the corn and wheat that is for the other farm animals.

Hawks can see really well. They can spot a mouse 100 feet away. That's a long long way!!! Out west they have lots of hawks and even bigger birds called eagles. I'll tell you about them in another story.

God is so smart!

He made such a beautiful world for us to study!!!

Hummingbirds

Today I want to tell you about the smallest bird God made, the hummingbird. People call them hummingbirds because they flap their wings so fast it sounds like they are humming. Can you hum? Try it.

Hummingbirds are very small. The smallest one is only two inches long. They can flap their wings up and down two hundred times in one second! See how many times you can flap your arms up and down in five seconds.

These birds have a long tongue that they use to lick the juice out of the middle of the flowers. They can lick their tongue out thirteen times every second! Try that!

Hummingbird eggs are super tiny. Their nest is so small, it would fit in your hand. They have really pretty feathers. It's almost like they glow. There is only one kind of hummingbird that lives in Florida, but there are 15 different kinds found in America and more than 300 kinds around the world.

The hummingbird is the only bird that can fly backwards! God is so smart! If "hummers" couldn't fly backwards, they would probably hurt the flowers while they get a drink. God thinks of everything!

One time when my children were your age, a hummingbird flew into our garage in Texas and couldn't find his way out. I caught him with a butterfly net and let the kids see how small and how pretty he was before we let him go outside.

When it starts to get cold in Florida, hummingbirds will fly south to Mexico or South America. It's hard to believe that those tiny birds can fly thousands of miles.

It will sure be fun to get to heaven and let God teach us about all the things He made. I love to learn new things! I loved school so much that I was even a teacher for a long time. I want all of you to enjoy school and enjoy learning to love God and His amazing world.

Kangaroos

I saw a picture of a kangaroo. God made 50 different kinds of kangaroos. The biggest ones are taller than me!

I saw some when I was in Australia. They can hop 40 miles per hour!

Kangaroos eat grass and leaves. Their babies are called joeys. They are only as big as a jellybean when they are born! They live in their mother's pouch and drink milk until they get big enough to go out and eat grass.

Kangaroos protect their babies! They can sit on their tails and kick really hard with their feet.

Don't mess with a baby joey unless you want to get kicked by the mother!

Maybe we can all go to Australia to see the kangaroos and tell the people there about Jesus.

There are some missionaries who live in Australia and tell the people there about Jesus now. We need to pray for them.

Can you find Australia on the globe without anyone showing you where it is? Find a map and look for a huge amount of land near New Zealand.

I love you!

Grandpa

Killdeers and Sandpipers

I love you, grandkids! I pray that you will grow up to love Jesus and serve Him all your life. When God came down to be born on earth, He was called Jesus. Jesus is God in a human body like we have.

God is s-o-o-o-o-o smart! I love studying all the things He made. Every night when I walk around the track and pray, I see a bird flying or running on the ground. He is called a killdeer. Killdeers are amazing birds! They build their nests on the ground. If another animal, like a fox, raccoon, or rat, comes toward their nest to eat the eggs or baby birds, the daddy bird will run away from the nest with one of wings hanging on the ground. He pretends like he is hurt. This makes the enemy run after him and away from the nest so the baby birds will be safe.

Like that daddy bird, everyone should protect the young ones that cannot protect themselves. You older children should protect your younger brothers and sisters.

When the daddy bird has made the enemies chase him far away from the nest, he jumps up and flies away. He tricks the bad guys to protect his family.

The killdeer is a type of bird called a plover. Most plovers fly a l-o-o-o-ng way every year to stay warm. They would freeze if they stayed in places where it snows in the winter.

One plover is called the Pacific Golden Plover. Like all of God's creatures, this one is amazing! The pacific Golden Plovers lay their eggs and hatch their babies in Alaska in the summer when there are millions of bugs to eat. Most birds love to eat bugs! Yuck!

After the babies get big enough to feed themselves, mom and dad fly away to Hawaii. They fly over 2,000 miles over the ocean where they have no place to stop, rest, eat, or sleep. It takes them about three days and nights of flying with no rest!

Two weeks later, the babies in Alaska are big enough to fly, so they fly to Hawaii to join their parents. How do they know the way to Hawaii? It's amazing that God can create this little bird with the map already in his brain to fly to a place he has never been!

Some plovers love to run along the edge of the water at the beach and eat little bugs or clams. Next time you go to the beach, see if you can see one. They are called sandpipers.

When the waves come in, they run up to stay out of the water, and then they run back down on the wet sand to eat before the next wave comes in.

There are also some killdeers that live around Dinosaur Adventure Land. They usually run around the field or parking lot, just about the time it gets dark. They make funny noises when they cry out to their family.   Can you find a bird like this?

Ladybugs

I love you grandkids! I also love Jesus and want to serve Him all my life. I pray that you will grow up to love and obey him, too.

The more I study God's world, the more amazed I am at how smart He is! Today I was walking around the track praying for people and I felt something on my left elbow. It sort of felt like a bug was biting me, but it didn't hurt like a bite. I looked down to see that it was a ladybug! She was hanging on so hard trying not to fall off, that it was pulling my skin!

Ladybugs are amazing little animals! They have huge wings they fold up under their shells. When they fly, they hold the shell up and unfold the wings.

Baby ladybugs are bright yellow and orange. Other bugs and birds don't eat them because they are poisonous. God gave them these bright colors to protect them.

Adult ladybugs can make a bad tasting juice come out of their skin so enemies won't eat them either.

Ladybugs are very helpful bugs. They eat tiny bugs, like the ones that kill orange trees. Next time you drink orange juice, you can thank a ladybug!

Ladybugs are also strong! God made them with 800 real tiny hairs on their legs so they can stick to glass and walk upside-down on the ceiling! They can hold five times their own weight! Imagine holding five of you on your back while walking on the ceiling!

God is s-o-o-o-o-o smart! I'm sure glad He lets me be in His family and is building a house for all of His children. If you have never asked Jesus to forgive your sins and live in your heart, you should do that today!

Opossums

A few nights ago, I was walking and saw a opossum (the "o" is silent, so it is called a possum)! He was a little bigger than a cat and sort of gray with a little white and black hair mixed in his coat. A opossum's face looks like a big mouse. Opossums eat plants, bugs, and small animals. They also eat garbage.

When mommy opossums have babies, they usually have six to ten at one time! When the babies are first born, they have to climb on their mother's fur to find milk and hang on tight because she just keeps climbing trees and walking in the woods looking for food. If a baby opossum falls off of his mommy, it is not strong enough or smart enough to find his mommy, so he dies.

Opossums hang by their tails and even sleep that way! When they get scared, they pretend they are dead or sleeping. That is why if you pretend that you are asleep, someone may say that you are "playing opossum."

The mother has a nice soft pouch of skin where the babies live. It is like a big pocket on her belly. The baby opossums live there until they are big enough to go hunt for food on their own. There are more than 65 different kinds of opossums in the world, but only one in the part of America where we live.

God is s-o-o-o-o smart! He made the baby opossum, which is only a little bigger than a raisin, know how to climb up on his mommy and find his milk as soon as he is born! Human babies can't climb up on their mommy as soon as they are born! Human mommies and daddies have to take care of their babies until they grow up.

Some people keep opossums for pets. I had one for a while, but they are not very smart, and they smell pretty bad. I decided to let mine go into the woods.

Don't try to touch them because they have sharp teeth and can bite hard! They aren't being mean, but they think that you are going to hurt them. If you are quiet when you walk at night, you may see one.

Owls

Last night I was walking around and praying for you when I heard an owl in the woods! They say "Whoooo! Whoooo!" I stopped and looked into the trees for a long time, but I couldn't see him, so I don't know which kind of owl it was. There are 18 different kinds of owls that live in America and 130 different kinds in the whole world.

The smallest owls only grow 5 inches tall and the biggest ones get 2 feet tall with wings that reach over 5 feet!

God is s-o-o-o-o-o smart! Everything He made about the owl is amazing. Owls hunt at night for rats and mice. Farmers love owls because rats and mice eat the farmer's corn and wheat. God gave owls special soft feathers on the front of their wings, so they make almost no noise when they fly. The mouse can hear really well, but he can't hear the owl coming until it's too late.

God also gave owls really special eyes! They are huge so they can see at night. Each eye is like a little telescope. They can see a mouse moving a l-o-o-o-ong way away. The owl cannot roll his eyes in his head like you can, so God also gave him a special neck. He can turn his head all the way backwards!

You can't do that! In the middle of the owl's eyes is a black spot, called the pupil, to let the light in. You have one in your eye, too. When it is dark, the black spot gets bigger to let more light in. The owl's pupil is huge.

Ask someone to close their eyes and count to ten slowly. Get up close and watch their pupils when they open their eyes. You have to watch carefully because it happens so fast, but when they first open their eyes, the pupil will be big. Then it gets smaller quickly because letting too much light in your eyes can hurt your eyes.

Did you know that owls sport different colored eyes for a reason? Those with dark brown or black eyes are only nocturnal hunters – they hunt at night. Owls with yellow eyes (snowy owls, great horned owls) hunt during the day, while those with orange eyes hunt during both day and night.

God is s-o-o-o-o-o smart. I can't wait to get to heaven and see all the things He has made for us up there!

Pigeons and Doves

Every day as I walk and talk with the Lord, I see the pigeons that live on the roof of my house. There is one that is pure white and is so pretty. Pigeons and doves are almost exactly the same. There are hundreds of different varieties of them and many have really pretty feathers.

Pigeons live all over the world. Lots of them live in big cities and eat food scraps that people drop. Many people like to hunt pigeons and doves to cook them for food. They taste really good!

Pigeons walk funny. Their head moves back and forth real fast. I don't see how they don't get a headache! Watch one walk sometime and see if you can do what they do.

Some pigeons live for 30 years! That's old for a bird. They are one of the only birds that can drink water without tipping their head back. Watch birds drink and you will see what I mean.

When mommy pigeons have baby pigeons, they usually have only one or two baby pigeons at a time. The babies are totally helpless. When the babies are hungry, they eat by putting their beak in their mother's mouth and eat out of that for their bowl! Yuck! I'm glad people don't eat like that!

Some pigeons are very good at finding their way home. They are called "homing pigeons." Uncle Ross used to raise some of these when we were boys. He could put some in a box where they couldn't see and drive them hundreds of miles away. When he opened the box, they would fly straight home! How do they know the way? Even though their brain is very small, God gave them a good one to be able to find their way home from anywhere!

Pigeons are mentioned in the Bible a lot. Noah let a dove out of the ark to see if the water was gone from the flood. The dove flew back to Noah with an olive leaf in its mouth (Genesis 8:11).

Doves are really sweet and soft. King Solomon wrote a song to his wife to tell her how much he loved her. He said she was like a dove (Song of Solomon 5:2)! When Jesus was baptized, the Holy Spirit of God came down on him like a dove (Matthew 3:1; Mark 1:10; Luke 3:12; John 1:32).

Next time you see a dove, it would be a good reminder to pray. God loves us s-o-o-o-o-o much! He always gives us things to remind us of His love.

Ravens

In Luke 12:24, Jesus told his disciples to "consider the ravens." S-o-o-o . . . let's consider the ravens like He told us to. Ravens are a big black bird like a crow. The word raven means black. They are mentioned eleven times in the Bible. Blue jays, crows, and ravens are all related.

The first thing to consider is that God made the raven when He made all the birds on day 5 of creation. Every creature God made is amazing and can teach us things about how smart God is! Just like every animal is different, every person is different too. God made you very special to do a special job for Him.

The raven can teach us some good lessons and some bad lessons. The first bird Noah let out of the ark was a raven. It never came back to him. Outside the ark were lots of dead rotting animals which ravens love to eat! YUCK! That's probably why God told His chosen people not to eat the raven (Leviticus 11:15; Deuteronomy 14:14).

God asked Job who provided food for the ravens and their babies. Job knew that God did this (Job 38:41)! Ravens are greedy birds. They will eat just about anything including other birds' eggs. There is only one time when ravens shared their food. God told the preacher Elijah to go hide by a brook because the wicked King Ahab was trying to kill him. While he hid by the brook, the ravens brought him bread and meat every morning and evening (I Kings 17). God knows how to make even the greedy ravens obey Him!

Ravens are smart birds but don't sing very prettily! They do watch out for each other. If an animal gets killed on the highway, the crows and ravens will eat it; but one bird is always up in a tree watching for danger.

The boy and girl crows and ravens look the same – all black. The daddy raven helps build the nest and feed the babies who eat a lot!

Some people have kept ravens as pets and taught them simple tricks such as how to say, "Cracker" like a parrot. Here are some things we can learn from the raven:

Ravens take care of each other – especially their family. That's good! We should take care of our families, too.

They love to eat dead rotting things in the world. It's good that they do that to clean up messes. We should clean up messes, too.

God told us to "consider the ravens." Ravens don't worry about food or clothes. They don't plant crops or build barns in order to save up lots of food. God provides for them (Luke 12:22). That's good! We should learn not to worry about food or clothes either. God provides for the ravens and we are much better than birds, so God will take care of us too (Luke 12:24)! Always serve God and obey His Word. He will always provide what you need!

Salmon

I love to eat salmon. It is one of my favorite dishes to eat. Yummy!

There are seven kinds of salmon in the world. Six of them live in the North Pacific Ocean and only one kind in the Atlantic Ocean. Ask someone to show you where those oceans are. 90% of the salmon we eat in America comes from Alaska. Maybe we can all go there one day and see them jump up waterfalls or even go fishing for some of our own salmon!

Every year millions of salmon swim up rivers and even jump up waterfalls! They go back where they were born to lay eggs. Bears always wait in the river to try and catch the salmon. Look for a video on the computer about salmon runs. You won't believe it!

Sloth

I'm reading lots of books trying to learn more. I love to learn new things about God's amazing creation. I pray that you will always love to learn too.

I saw this picture of a sloth and decided to read all about sloths. There are two kinds of sloths. Both kinds have three toes on their back feet, but one kind has three toes on its front feet, while the other only has two toes on its front feet.

Both kinds come from South American or Central America and live in trees. They hang upside down all the time! They even sleep hanging by their toes. They move *really slow*. The only time they come down out of the trees is one time each week to go potty. It takes them over an hour to *slowly* climb down the tree and back up again.

Sloths have long smelly hair. Lots of them have green mold growing on their hair so they are hard to see in the trees.

People who are lazy or really slow are called slothful. The word slothful is used seventeen times in the Bible. God does not want us to be slothful! We should always work hard at every job we do.

Maybe you can go to South America and see a sloth some day! That would be fun!!!

Sparrows

I got to feed the sparrow again today. It is a mommy sparrow. The daddy sparrows have black feathers under their chins. These black feathers make it look like the daddy sparrow has on a bib!

This mommy sparrow is not afraid of people! She comes right down on the sidewalk by me and I throw sunflower seeds to her. She is really smart, too. She flies up under the awning and scares the bugs out. When the bugs start flying out of the awning, she catches them right out of the air.

She works hard to get lots of food so I think she must have some babies to feed. Right now, I see that she brought two other mommy sparrows with her. She probably told them about the free food.

Birds can't sweat to cool off like we do. Instead they hold their beak open and breathe out the hot air to cool off. Three of them are sitting outside the window now with their mouths open because it is hot outside. Now I see twenty-four sparrows. It's fun to throw bread out for them. I love watching birds. God is s-o-o-o-o smart.

There is a very tiny animal called a mite that likes to live in the feathers of the sparrow. It makes a tiny bite into the sparrow to get some blood like a mosquito does. The sparrow doesn't like them, but birds don't have bug spray, so they all find some dust to roll around in. This puts a fine dust all over the feathers to keep the mites off. We call it a "dust bath." Imagine taking a bath to try to get dirty! Pigs, elephants, and hippos cover themselves with mud to keep the bugs off their skin too.

Sparrows may be small, but they fight hard to protect their babies! Crows like to try to steal the sparrows' eggs, but the sparrows will fly at the crow to fight them off. They even pull feathers out of the crow! Next time you see a crow fly overhead, look to see if he is missing some feathers. If he is, he probably got beat up by some small sparrows! Maybe that will teach the crows to leave the sparrow's eggs alone!

Maybe you can find a sparrow's nest and see the babies! Don't get too close to them or touch them though, or the mommy sparrow will get really mad at you! She will think you are trying to hurt her babies and she will try to hurt you!

The Bible says the sparrow loves to make its nest in God's house (Psalm 84:3). The next time you go to church, look for sparrows. The Bible also says that God knows all the sparrows and He loves us even more than the sparrow (Matthew 10:29; Luke 12:6). I am so thankful that God knows where we are and watches over each of us!

Squirrels

As I was walking, I saw a squirrel today, so I'll tell you about squirrels. God must really like them because He sure made a lot.

There are 260 different kinds of squirrels found almost all over the world. Some live in the ground and some live in trees. Some make a barking noise, like ours, but others whistle or chirp.

There are some tiny flying squirrels that live in the trees but they are real shy and only come out at night so you probably won't see them. They glide from branch to branch; they can't really fly.

God is s-o-o-o-o smart! Squirrels chew on nuts and tree branches all the time. This wears out their front teeth so God made it where their four front teeth grow all their life!

Squirrels' tails are amazing too. They use them for balance when they run or climb trees, and for shade when it gets hot. It's like they carry their own umbrellas!

I pray that you always love Jesus and love to study His amazing world.

Turtles

I have a lot of fun learning about God's creation. I love to read books and learn new things. Did you know God made a really amazing animal when He made a turtle? There are thirteen basic kinds of turtles. Some live on land, some live on both land and water, and some only live in the water – they are the biggest. Some of the water turtles get bigger than your kitchen table!

All turtles hatch from eggs that the mother lays on land in a hole she digs for them. When they hatch out they are on their own to find food. Their mommies don't feed them when they are babies. God made them so they just know what to do as soon as they hatch!

Sea turtles have flippers so they can swim.

Lots of turtles have germs on them that will make you sick. If you touch a turtle, be sure to wash your hands with soap before you touch your mouth!

Turtles don't have teeth, but some can still bite hard. The snapping turtle can bite right through a broom stick! He could bite off your finger!

We can learn a lot from turtles. They can't talk or make any noise. Like them, we need to learn to be quiet and let God talk to us. That is called a quiet time.

Turtles can't run fast, so they learn to hide inside their shells for protection. We can't fight the devil; so we need to hide in Jesus and let Him protect us.

Turtles never get in a big hurry and they live a long time. We need to learn to relax, chill out, and enjoy God's wonderful world.

I pray that you will all grow up to love and obey Jesus all your life. Just like God made turtles special, He made each person special too. You should learn how to do as many things as you can, so that Jesus can use you anywhere.

Whippoorwills

When it gets dark, I like to walk around and talk to the Lord. I also like to listen to the birds that come out at night. Some birds sleep at night and fly around all day, but other birds sleep all day and fly at night. We have two birds I hear every night. They call to each other. They are called whippoorwills. They sing the same song over and over all night, which gets a little bit annoying! They love to fly around and eat bugs.

God is s-o-o-o-o-o smart! Some bugs sleep at night and are awake during the day. The day birds can see them moving around and eat them for food. The night birds see the night bugs moving around and eat them. If birds didn't eat bugs, the bugs would fill up the whole world and we couldn't go outside!

The whippoorwill is the only bird that sleeps all winter when it is cold. That is called "hibernating." Since most bugs also sleep all winter, God made the whippoorwill sleep too since they would not be able to find food.

God loves all of His creation. He especially loves you!

I pray that you will always love Him and love telling others about Him and His amazing world.

Canadian Geese

For the last few days, big Canadian geese have been flying overhead. They fly in a "V" shape because each one gets a little bit of extra lift from the wings of the one ahead of it. The one in front doesn't get any help, so when he gets tired, he lets a different one lead the flock. They all take turns doing the hard job and never complain. We should do that too!

Then, yesterday, about twenty of the geese landed in the grass right by the fence to eat bugs and rest up. It was s-o-o-o-o cool to see them all up close like that. They are big!

Don't get to close to these birds. They might think you are going to hurt them, and they can bite hard and hit you with their huge wings. They are not mean, but they will protect themselves.

Cats

Today I want to tell you about cats. They are a really amazing animal! God must like them because He made s-o-o-o-o-o many different kinds of colors.

Some cats, like lions and tigers, are huge! Some cats have spots, some have stripes, and some are all just one color. Almost all of them have regular tails but a few cats have short tails. See if you can find them in the pictures.

There are many amazing things to learn about cats. When a cat moves

her babies to a new house, she picks each one up in her mouth and carries the baby by the neck! It doesn't even hurt them!

The cat's eye is really neat. When the sun is out, the black spot in his eye gets small like a line. Yours gets smaller in a circle. When it's dark, the black spot opens up bigger to let more light in. That's why cats can see really well at night when it is almost dark.

Their tongue has hundreds of tiny hooks. They comb their hair with their tongue! Yuck! I'm glad we don't do that! When a cat licks you, it feels rough like sandpaper, but it won't hurt you. Put some milk on your hand and let a cat lick it off so you can feel their special tongue. God thinks of everything! He loves you! So do I!

Cats 2

Dear Grandkids,

God made lots of kinds of cats; more than forty different kinds! The lions and tigers are the biggest kinds. The ones people keep as pets are called house cats.

There are almost forty kinds of house cats. All cats can purr and no one knows why they do it. All cats, except the cheetah, can pull their claws in so they don't scratch you.

The cats' eye is amazing too. The black part in our eye is always round, but cats' eyes close to a black slit when it's bright out and open up to a big black circle when it's dark. They can see real well at night so they can catch mice and rats. Some cats, like the manx, don't have a tail. Some cats have long hair and some have short hair. Cats also have a very special tongue that combs their hair!

Cats make very good pets, but they are pretty independent and hard to train. Some of the people in Egypt a l-o-o-o-o-o-n-g time ago thought cats were gods! That idea is s-o-o-o-o-o-o dumb! God made the cats like He made all the other animals. It would be silly to pray to a cat. Cats can't answer your prayers. Only the real God can do that!

I pray that you will always grow to love and obey Jesus.

Cockroaches

We can all take walks and squash cockroaches! God is s-o-o-o-o smart! He made almost 4,500 different kinds of cockroaches! They live all over the world. They don't bite, but they eat anything. They especially love garbage and rotting food.

When cockroaches walk on the garbage, they get lots of bacteria on their feet. Then they walk on your table or food and the bacteria gets on your food. Then when you eat, the bacteria can get in you and make you sick.

The mommy cockroach will have about forty babies at a time! She lays eggs and carries them with her for a while until she finds a nice place to leave them so they can hatch.

Cockroaches love to eat dead leaves around trees. God made them for that job to make sure the leaves wouldn't get too deep and hurt the trees. Most cockroaches can fly, but they would rather run around at night so birds can't see them and eat them.

God thinks of everything! I can't wait till we all go live in God's huge house in heaven!

Crickets

Every time I study things that God made, I am more amazed at how smart He is! If you study how bugs make noise, it will make you praise God for His wisdom.

We make music many different ways like using drums, pianos, and trumpets. Insects have neat ways to make music. When male crickets want to find their family at night in the dark, they rub their wings together really fast. The edges of the wings (the scrapers) rubbing together is what makes the chirping sound. Females do not chirp.

Put a credit card on the edge of a table halfway off and rub a comb over it. It will make a sound sort of like a cricket does with his wings. A file or wood rasp will make a different sound.

Some insects rub their legs to make this sound. Every cricket makes a different sound to find his family. They all sound the same to me! How do the crickets know which cricket is calling them?

Crickets' ears are on their elbows! I'm glad God put our ears on our head! Other insects, like locust, have hollow spots inside their bodies and they move tiny muscles really fast to make a sound like a tiny drum, but it beats really fast. Most of them chirp faster when it gets warmer.

If God can make tiny insects that are s-o-o-o-o-o- cool, imagine what kind of house He has built for us in heaven! I can't wait to go see it!

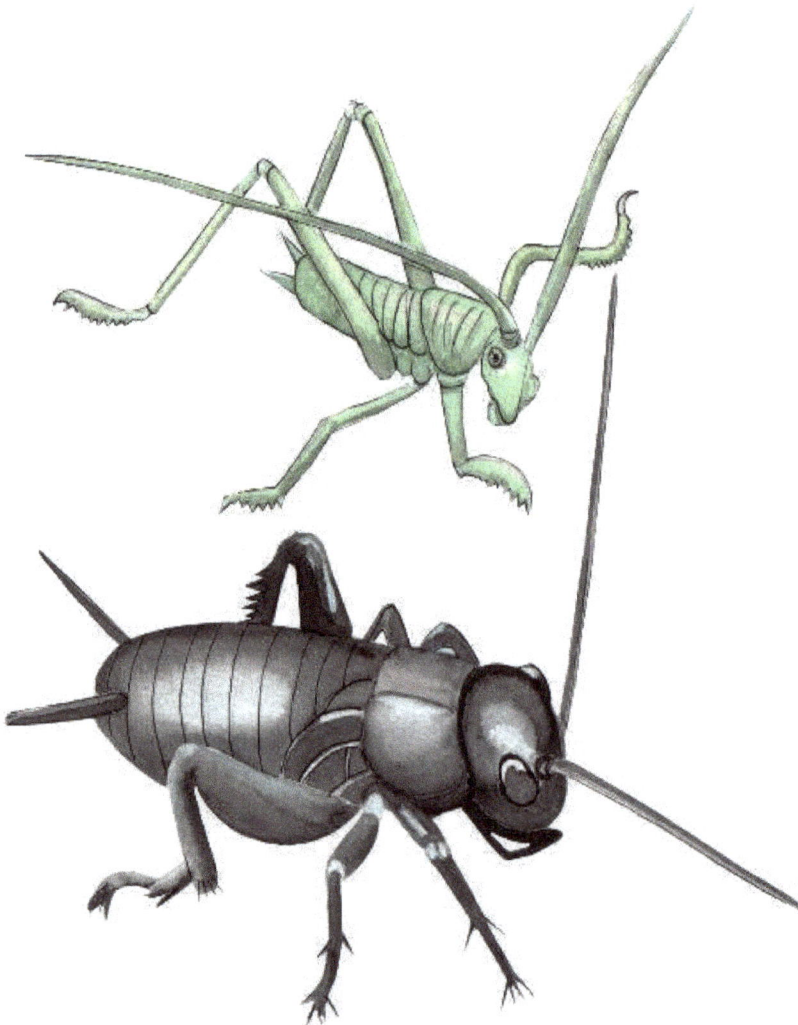

Crows

Today I found a crow feather.

A crow is like a raven.

The crow probably tried to steal some food from a small bird and the small bird pulled out one of its feathers. Now he will have a hard time flying for a while until his missing feather grows back.

This crow might learn to be nice next time!!

Grandpa

Deer

Tonight I looked out the window and saw deer eating grass. They were s-o-o-o-o-o pretty! God made amazing animals. There are about thirty-five different kinds of deer in the world. The deer I saw today were all white-tail deer.

The smallest deer only gets one foot tall while the biggest gets almost eight feet tall! Almost all of the boy deer grow antlers every year. The older they get, the more antlers they grow.

The baby deer is called a fawn. They have spots all over to help them hide. They can walk in less than an hour after they are born! Human babies can't walk for almost a year.

Some people train deer to pull sleds like a horse.

Deer love to eat grass, leaves, and bark off trees. Most deer can't make any sounds, but they can run forty miles per hour and can swim well too.

Deer meat is good to eat and their skins make nice leather for shoes, purses, coats, and belts. We have some deer skins and antlers at Dinosaur Adventure Land.

Fawns

This picture of two fawns made me think of you. One of them was just born and can already stand up!

God is s-o-o-o-o-o smart! He made fawns with spots so they are hard to see if they hold very still. Their mom cannot talk to teach them, but fawns know to hold still if an enemy comes close. The mom will run away from the baby so the enemy will chase her and leave the baby alone. Fawns can run just a few hours after they are born. Human babies can't even walk for almost a year!

Almost all animals have a smell that their enemies can smell but fawns have no smell! Who do you think made them that way? I know! God did!

When God fixes the earth back like it used to be when He first made it, all the animals will be friendly again. We will be able to go play with the deer. I want to pet a fawn like this! May we could have one for a pet! That would be s-o-o-o-o cool!

I love studying God's amazing world with you. He made so many wonderful plants, animals, and rocks that we could never learn everything about all of them. But . . . it is fun to try!

Today, let's study the fly. There are 120,000 varieties of flies! We will only talk about the one called the "house fly" because they are the ones you are most likely to find in your house. Flies don't try to hurt you, but they do because they love to eat garbage and then walk on your food! Yuck! This puts germs from the fly's feet and tongue on your food and it can make you sick.

Flies start off as a tiny egg. After one day, they hatch into a little worm called a maggot. Maggots love to eat rotting stinking food! Yuck! After five days they turn into a worm-like creature called a "pupa" covered by a hard shell like a caterpillar's cocoon. Inside the shell the worm-like pupa turns into a fly! It is amazing how this happens. In two weeks the fly has its own babies!

Flies smell food with their antennae but they taste it with their feet! Then they taste it again with the hairs on their lip. The best way to keep flies out of your house is to keep food off the floor and table. Also take the trash out often. Flies love trash! Always keep everything clean and the flies will have nothing to eat and will stay away.

Flies have two large eyes that can see in almost all directions at the same time! They also have three more small eyes between the two big ones. You would look funny with five eyes!

Some places have big black flies that bite horses and cows and

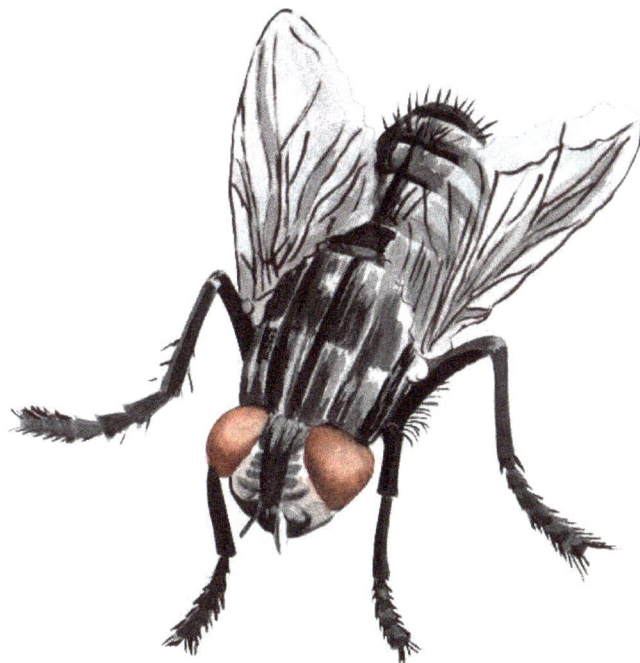

drive them crazy. There is another fly that loves fruit and it ruins many kinds of fruits that people like to eat.

Many animals, like birds, frogs, and dragonflies, love to eat flies. It would take 1,000 flies to weigh as much as six pieces of paper which is about one ounce. We'll talk about dragonflies another time.

Obey your parents. Love Jesus and learn to read so you can read the Bible and help others know about Jesus. Love, Grandpa

Foxes

Tonight, I got to feed the raccoon and the fox! There are two kinds of foxes in America, the red fox and the black or silver fox. The one we have here is a black fox. He is mostly silver with black at the tip of his tail. I don't know why they call them a black fox when they are really silver colored!

Our fox is only a little bigger than a cat with a long very bushy tail. Foxes are like small dogs, but they can run superfast. Samson in the Bible caught three-hundred foxes at one time. He must have been really fast too! Jesus talked about foxes in Matthew 8:20 and in Luke 9:58 and Luke 13:32.

Foxes eat small animals like mice, rats, rabbits, and frogs. Tonight the fox ate bread from me. They have soft fur to keep them warm. We have a real fox skin in the DAL museum. Come and feel how soft it is. God is s-o-o-o-o smart! He made many amazing animals. I love to learn about them. Someday Jesus will come back to rule the world and all the animals will stop being afraid of us. Then we will be able to play with them.

Keep serving Jesus!

Frogs

I sure love you! I pray for you to grow up to be in love with Jesus and work for His kingdom. God is s-o-o-o-o smart! Everything that He made in one second, we can study for thousands of years and never understand it all!

There is a really tiny pond right by the fence here. There are at least three big frogs that live in it. They are called bull frogs. I was reading a book about frogs and I learned a lot! Frogs are different from toads. Toads don't live in the water like frogs do. Toads have bumpy skin and frogs have smooth skin.

God made over four thousand kinds of frogs!   The littlest one is only as big as your fingernail! That is as big as it gets! The biggest frog is as big as a small cat! Frogs and toads love to eat bugs. Yuck! It's a good thing they do, or else there would be so many bugs that we wouldn't be able to go outside. One frog can eat hundreds of bugs every day. God is really smart. He thinks of everything!

He thinks of everything for you, too. He knows every hair on your head and everything you even think about. He loves you s-o-o-o-o-o much! He wants you to love Him, but you don't have to. He lets you choose. I love God and I want to work for Him all my life.

It is my prayer that these stories will draw you, your children, and grandchildren closer to the great God of the universe.

Goslings 2 – Learning to Fly

Today I saw something that was s-o-o-o amazing! I just have to tell you about it. The ten baby geese that were hatched out of their eggs just five months ago are as big as their moms and dads already!

I couldn't believe it. In only five months they grew huge. Every day they have been following their mom and dad around learning what to eat and how to watch out for their enemies like the hawk or the fox. Every day their mom and dad taught them how to swim and find food in the water.

Today, they were learning how to fly like a goose. They flew over us and honked their funny honking sound. The parents were trying to show them how to fly in a V formation to save energy. If geese don't learn how to fly in the right spot in the big V, they will get too worn out to fly south in the winter.

When the front goose flaps his wings it makes the air spin behind the sides of his wing tips. If the next goose is in just the right spot, the spinning air helps him fly lots easier.

They were practicing right over our heads! If each goose does his job and flies in just the right spot, all the geese have it better. It's easier to fly for all of them. That's why it's always best for us to do our job in the family. It makes everyone's jobs easier!

It was s-o-o-o-o cool to see the mom and dad teach the babies about flying. Geese don't have as many things to learn like people do. Wow! Geese learn to grow up in only five months, but we humans have so much to learn from our parents that God made us grow up *lots* slower. It takes people about 20 years to learn enough to be on their own and all grown up.

I'm so glad God gave you good mommies and daddies that want to teach you things. Be sure to listen to them and learn to do your job in the family. If you have Jesus in your heart you are one of God's kids. Be sure to read your bible to learn what He wants you to do.

Keep watching the sky in the fall to see if the geese fly over your house. You will see them flying in a huge V shape. God is s-o-o-o-o smart! I love studying His amazing creation!

For more information on the subject of creation, evolution, and dinosaurs, see our award winning DVD series. You can get your copy here:

Creation Science Evangelism
488 Pearl Lane
Repton, AL 36475
www.drdino.com

The Gospel

Even better than knowing about creation is knowing The Creator!
God wants you to know Him. He loves you and wants you to become His child.
How can you be His child when He is perfect and you are not? Good question!
It's as simple as A, B, C!

A - All have sinned. (Romans 3:23)
B - Believe on Jesus who died for you. (Romans 5:8)
C - Confess your sin (Romans 10: 9, 13)

Sample Sinner's Prayer

Dear Lord Jesus,
I'm a sinner.
I believe you died for me on the cross and rose from the dead.
Will you please forgive me, move into my heart, and save me right now?
Thank you Jesus!
Amen

If you do receive Jesus, that makes you
one of God's children
(John 1:12 & John 3:3-19)

Please call and tell me
so I can rejoice with you!
855.BIG.DINO (244.3466)

Grandpa Hovind

www.ingramcontent.com/pod-product-compliance
Lightning Source LLC
Chambersburg PA
CBHW052346210326

41597CB00037B/6270

9 781733 512855